Date: 12/6/21

**J 534 DAH
Dahl, Michael,
Sound waves /**

PALM BEACH COUNTY
LIBRARY SYSTEM
3650 SUMMIT BLVD.
WEST PALM BEACH, FL 33406

WAVES IN MOTION

SOUND WAVES

by Michael Dahl

PEBBLE
a capstone imprint

Pebble Emerge is published by Pebble, an imprint of Capstone.
1710 Roe Crest Drive
North Mankato, Minnesota 56003
www.capstonepub.com

Copyright © 2021 by Capstone. All rights reserved. No part of this publication may be reproduced in whole or in part, or stored in a retrieval system, or transmitted in any form or by any means, electronic, mechanical, photocopying, recording, or otherwise, without written permission of the publisher.

Library of Congress Cataloging-in-Publication Data is available on the Library of Congress website.
ISBN: 978-1-9771-2274-2 (library binding)
ISBN: 978-1-9771-2619-1 (paperback)
ISBN: 978-1-9771-2301-5 (eBook PDF)

Summary: We hear sound all around us. But how? Learn about sound waves and how we use them every day.

Image Credits
Capstone Studio: Karon Dubke, 20; Shutterstock: Andrea Danti, 8, Cristi Matei, 11, Dean Drobot, 14, Firefighter Montreal, 19, gritsalak karalak, 15, I AM CONTRIBUTOR, 7, ilusmedical, 9, Paul Reeves Photography, 13, Rudmer Zwerver, 17, Sergey Novikov, 5, Steve Byland, Cover, STILLFX, 12

Design Elements
Capstone; Shutterstock: Miloje, Ursa Major

Editorial Credits
Editor: Michelle Parkin; Designer: Ted Williams; Media Researcher: Jo Miller; Production Specialist: Laura Manthe

All internet sites appearing in back matter were available and accurate when this book was sent to press.

Printed and bound in China.
3322

TABLE OF CONTENTS

WAVES 4

MOVING THE AIR 6

YOUR EAR 8

SOUND MESSAGES 10

PITCH 12

ECHO 14

SOUND ALL AROUND 18

MAKE MUSICAL SOUND WAVES 20

    GLOSSARY 22

    READ MORE 23

    INTERNET SITES 23

    INDEX 24

Words in **bold** are in the glossary.

WAVES

Have you ever been to the beach? Have you watched the **waves** move through the water? The waves move all around.

There are other types of waves that you cannot see. These waves are all around us. They move through the air. They are called sound waves. Sound waves help our ears hear.

MOVING THE AIR

Sound waves are made by moving air. Have a friend hold a rubber band. Tell your friend to hold the ends tight. Pull the middle of the rubber band and let it go!

Watch how the rubber band moves back and forth. The moving band is **vibrating**. The vibrations move the air around it. Do you hear the sound it makes?

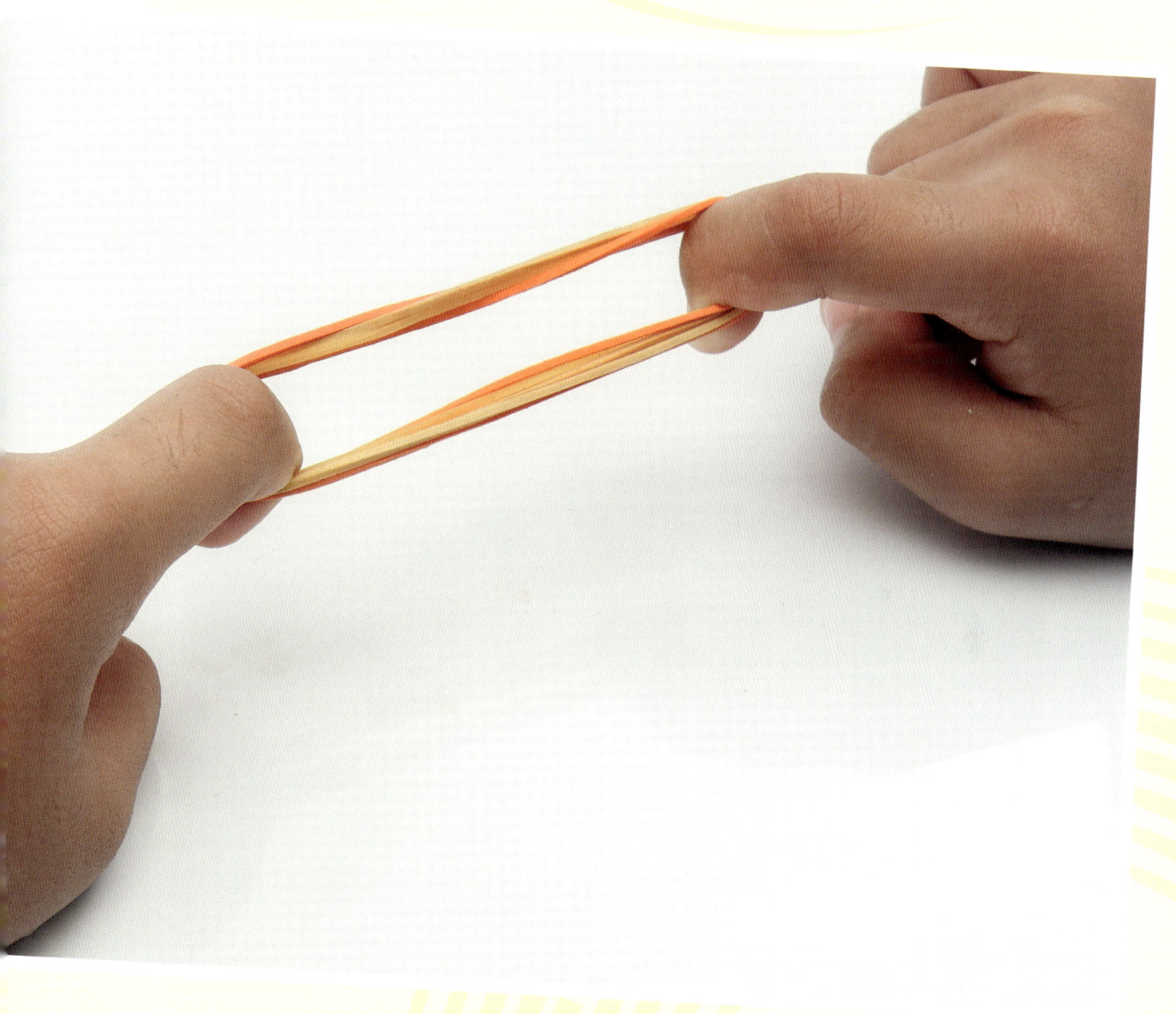

YOUR EAR

You cannot see a sound wave. You can only hear it. When the sound wave reaches your ear, it goes inside.

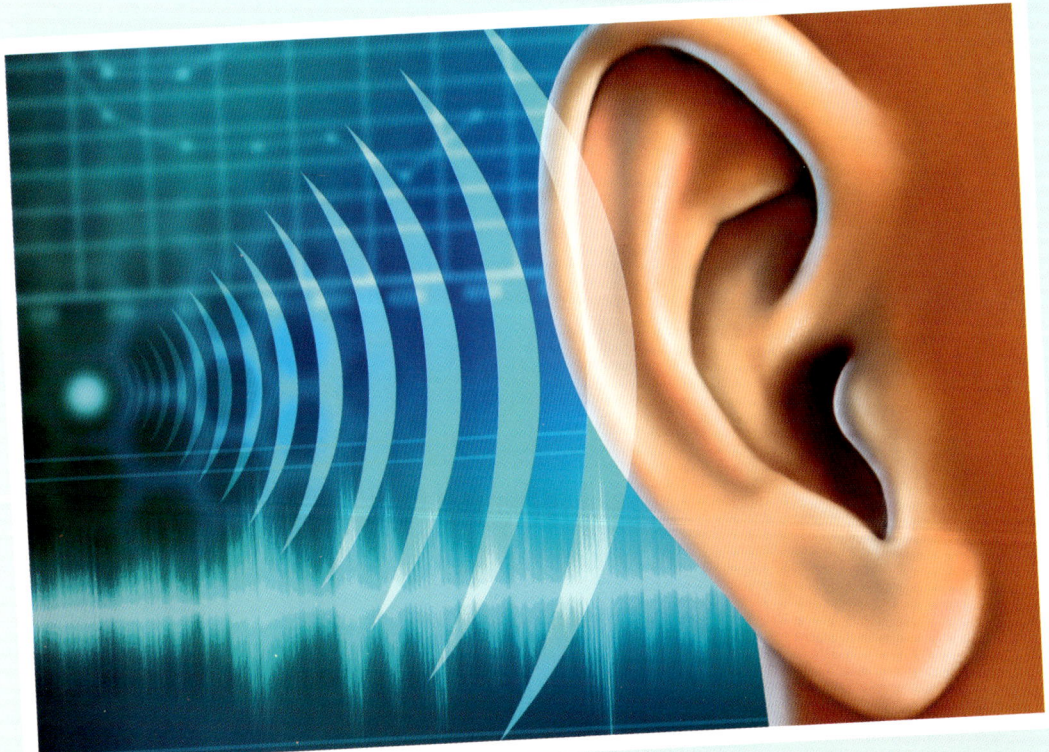

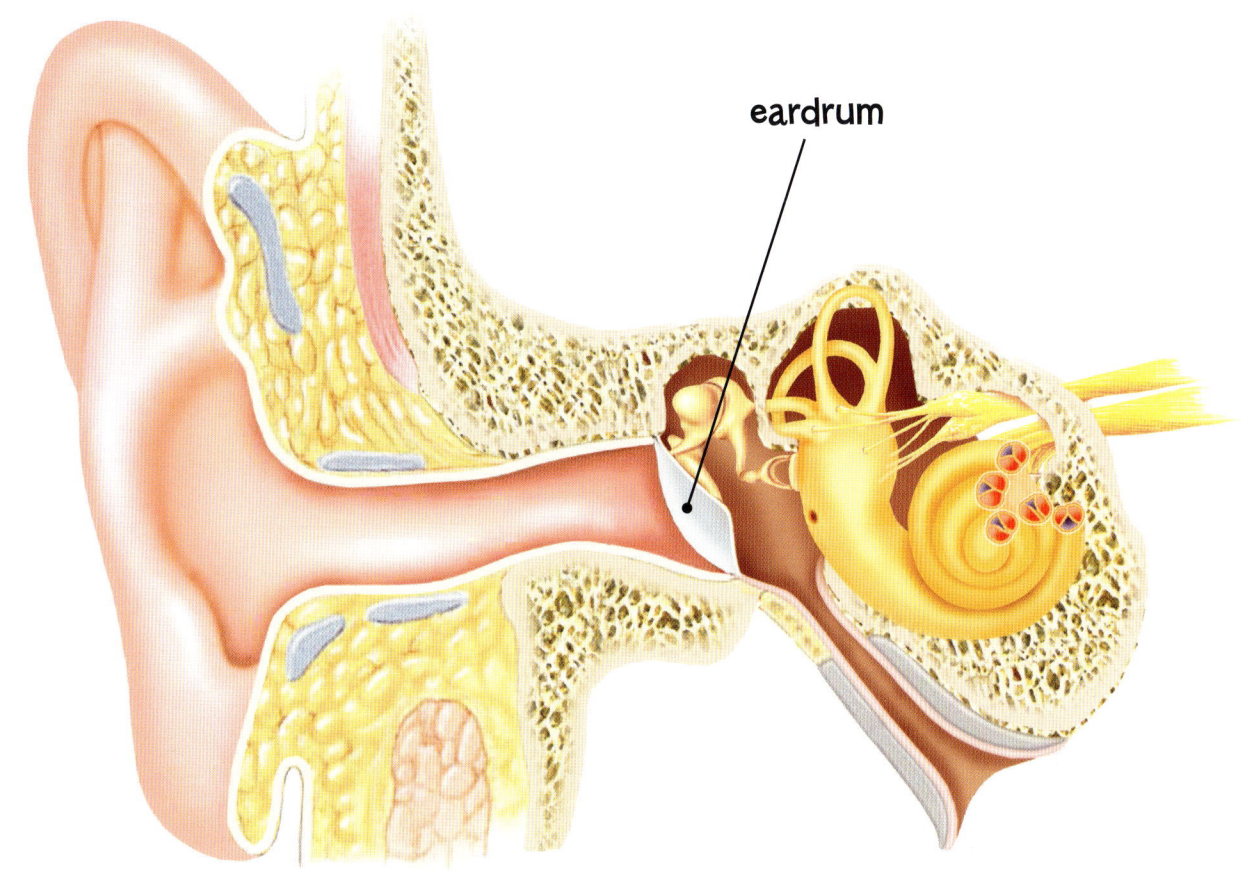

The **eardrum** is deep inside your ear. The moving sound wave touches the eardrum. The eardrum starts to move back and forth. The eardrum vibrates.

SOUND MESSAGES

When your eardrum vibrates, it sends a message to your brain. Your brain tells you what the sound is. Sounds can be loud like a balloon popping or soft like a whisper.

You brain can tell where sound waves come from. When a dog barks, you don't have to see the dog to know where it is. Your brain tells you.

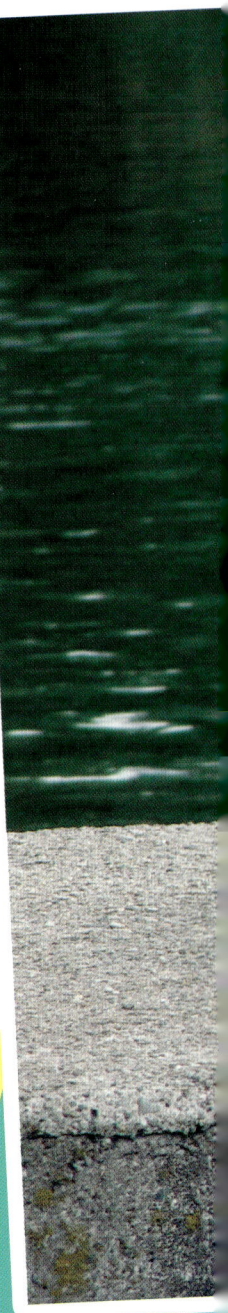

PITCH

Remember the moving rubber band? A thin band will make a high sound. A thicker band will make a low sound. **Pitch** is the highness or lowness of a sound.

A whistle makes a high sound. A rumble of thunder is a low sound. Listen to a bird sing. Listen to a dog growl. Which sound is high? Which sound is low?

ECHO

Sound waves can bounce. When waves hit something hard, they bounce back.

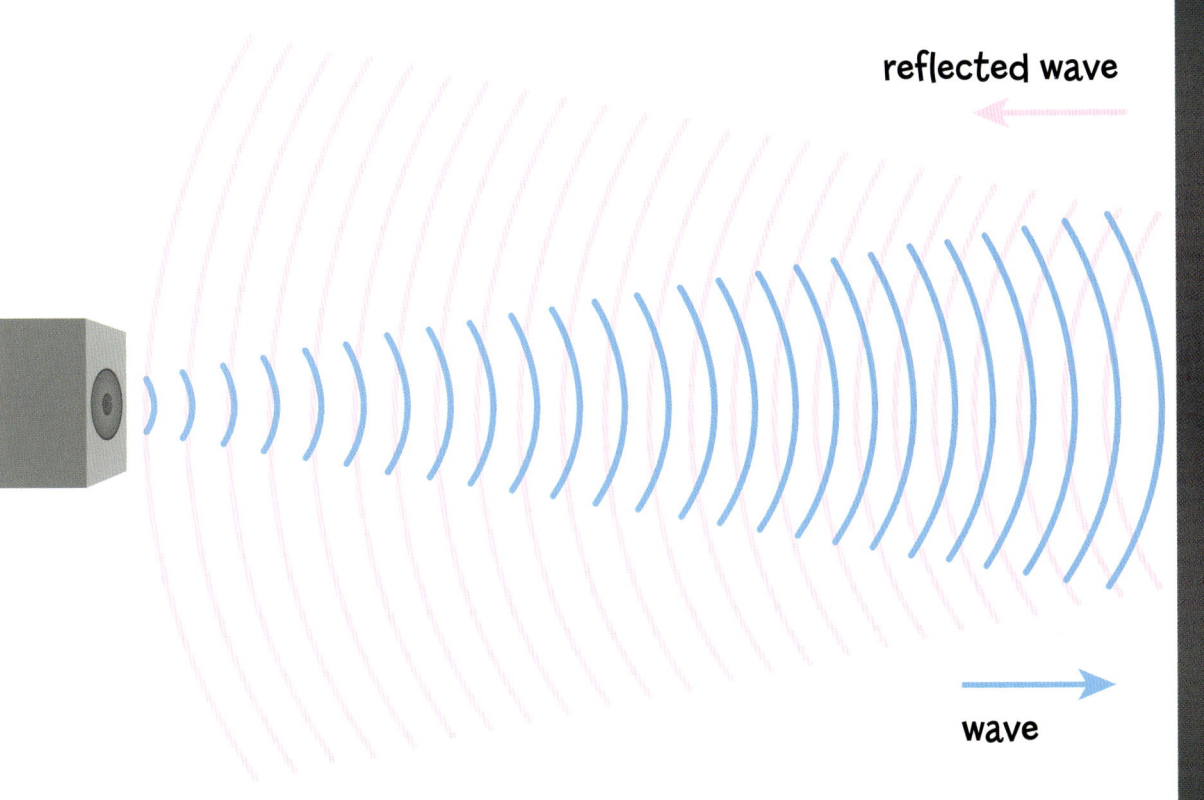

Shout "hello" in an empty room. Sound waves leave your mouth and hit the walls. The waves bounce back fast. You may hear the word again. This is called an **echo**.

Bats use bouncing waves to help them fly. Bats can't see very well. But they have very good ears.

The bat calls out. The sound wave moves fast. The wave hits a tree. The sound bounces back. The bat hears the echo. This tells the bat that something is in its way. Now the bat can safely fly around the tree.

SOUND ALL AROUND

Sound waves tell us information. We know class is starting when we hear the school bell. We know a fire truck is coming when we hear its siren.

When you finish this book, make a sound. Close the book's cover slowly. Make a sound as quiet as you can. Shhhhhhhh!

MAKE MUSICAL SOUND WAVES

Find out if different amounts of water make different sounds.

What You Need:

- three or four water glasses (made of glass)
- a metal spoon or knife

What You Do:
1. Place the glasses in a line.
2. Fill the glasses with different amounts of water.
3. Use a metal spoon or knife to gently tap the side of each glass.
4. Listen carefully.

Can you hear a different pitch from each glass? Are some sounds higher or lower than others?

Change the water amount in the glasses. Does that change the pitch? Why or why not?

GLOSSARY

eardrum (IHR-drum)—a part of your ear that helps you hear sound waves

echo (EK-oh)—a sound wave that bounces back to your ears

pitch (PITCH)—how high or low a sound is

vibrate (VY-brate)—to move back and forth quickly

wave (WAYV)—energy that moves through water or air

READ MORE

Diehn, Andi. *Waves: Physical Science for Kids.* White River Junction, VT: Nomad Press, 2018.

Gregory, Josh. *Sound.* NY: Childrens's Press, 2019.

James, Emily. *The Simple Science of Sound.* North Mankato, MN: Capstone Press, 2018.

INTERNET SITES

Sound
https://www.dkfindout.com/us/science/sound/

Sound Waves!
https://www.funkidslive.com/learn/waves/sound-waves/

INDEX

air, 4, 6

bats, 16

brain, 10

eardrums, 9, 10

ears, 4, 8, 9

echoes, 15, 16

hearing, 4, 6, 8, 10, 15, 18

loud sounds, 10

pitch, 12

soft sounds, 10

sound waves, 4, 6, 8, 9, 10, 14, 15, 16, 18

vibrating, 6, 9, 10